A Pissis

Nitrate and guano deposits in the desert of Atacama

An account of the measures taken by the government of Chile to facilitate the development thereof

A Pissis

Nitrate and guano deposits in the desert of Atacama
An account of the measures taken by the government of Chile to facilitate the development thereof

ISBN/EAN: 9783742863249

Manufactured in Europe, USA, Canada, Australia, Japa

Cover: Foto ©ninafisch / pixelio.de

Manufactured and distributed by brebook publishing software (www.brebook.com)

A Pissis

Nitrate and guano deposits in the desert of Atacama

INTRODUCTION.

For some years past the rich deposits of Nitrate of Soda, and Guano, as also the Mines of Silver and Copper which exist in the northern extremity of the Chilian territory in the desert of Atacama, have excited much interest. Some bold explorers have undertaken and accomplished journeys of investigation, and sanguine capitalists have established enterprises for their development.

But all attempts made by private individuals have met with considerable obstacles. The land is barren; supplies of potable water are frequently unattainable; it is cut off from the sea by the coast-range of the Cordillera, and the existing means of communication are both sparse and costly.

The Chilian Government, desirous of rendering aid to private enterprise, has sent thither two Commissioners, one for the purpose of discovering the natural sources of wealth existing in this desert, until now hidden from view, and the other to study the means of affording ready access for those who may be engaged in the work of bringing its produce within reach of the markets of the world.

The object of this pamphlet is to place under the eye of the public the information thus obtained by publishing the text itself of the Commissioners' Reports. The most important of these documents is undoubtedly the report of Monsieur Pissis, which establishes the fact of the existence of abundant natural riches in the desert of Atacama.

The deposits of Nitrate of Soda which have been recognized are undoubtedly of considerable extent; and it may be assumed, with much probability, that these constitute but a small part of the existing deposits of the entire desert. As much may be said of the Guanos, and of the minerals—Silver and Copper.

It was impossible, in so short a time, to effect a complete exploration of the whole of the riches of this desert, whilst for the present it is sufficient to indicate the prominent points of attack; when once exploratory works have been organized, the exploration may with little trouble be extended step by step.

The Commission charged to select a place where the works established in the desert may, at the least cost and labour, be placed in communication with the sea, for importation of stores and export of produce, has dis-

INTRODUCTION. iii

covered a very convenient port capable of being connected with the interior by a cart-road in a short space of time. Such are the most essential and pressing matters. Time will do the rest.

The Chilian Government is desirous of taking advantage as soon as possible of the studies of the Commission, by undertaking immediately the construction of the port-works of Blanca, Encalado, and Taltal. Those for the construction of roads will follow hereafter.

In this way capitalists, engineers, and contractors who may be disposed to extend their operations to the borderr of the Pacific, in view of increasing the exportation of Nitrate of Soda and other products, will be in possession of positive and reliable information. It is to these persons that the Government of Chile addresses the present publication.

THE NITRATES AND GUANOS
OF THE
DESERT OF ATACAMA.

[Translation.]

No. I.

Santiago, 30th June, 1877.

To His Excellency the Minister of the Interior, Don José Victorino Lasturria.

Sir,

In the execution of the commission with which you have entrusted me, I have the honour to submit to you herewith a Report upon the Geology and the Mineral Products of the Desert of Atacama.

I am, &c.

(Signed) A PISSIS.

REPORT

UPON THE

DESERT OF ATACAMA,

ITS GEOLOGY AND MINERAL PRODUCTS.

1. GENERAL ASPECT.

THE desert of Atacama is generally regarded as an extensive plain enclosed between two chains of mountains, the coast-range on the west, and the Andes on the east.

This description is very far from being correct.

The interior of the desert does not consist wholly of level plains, but is divided into large basins by intersecting ridges, the direction of which is approximately north-east to south-west.

Between the parallels 23° and 27° there are four of these basins wherein may still be seen the dry channels of the ancient rivers which once irrigated these extensive regions. The northern basin embraces the whole of the space enclosed by the hills of Naguayan, Caracoles, and Atacama on the north. Its eastern boundary consists of part of the Andes range, which extends from the volcano of Licancaur to Llullaillaco. Its southern boundary is a range which, branching from the Varas mountain, runs in the direction of the heights of Los Cordones and Cobre, and terminates on the coast by the hills of Jara and Jorgillo. This vast basin communicates with the Pacific Ocean through a deep gorge called the Negra, in the vicinity of Antofagasta. The portion of this basin which is in Chilian territory comprises the plains occurring in latitude 24°, namely, those of Palestina, of Aguas Blancas, and a part of the valley of Mateo.

The second basin, that of Cachiynyal, is bounded on the north by the peak of Cobre and by the Varas chain, on the east by the Vaguilla range as far as Chaco, thence by the Andes as far as the volcano of Doña Ines, and on the south by a range of hills which includes those of Hornillo and Cachiynyal. This basin opens into the port of Taltal, and contains the most extensive tracts of level ground in the desert of Chile, viz., those of the Prophète, Cachinal, Sandon, the valley of the Encantada, and the plains of Cachiynyal.

The third basin is of more limited extent, and contains but few tracts of level ground; it is bounded on the north by the range last indicated, on the east by the volcano Doña Ines and the hill of Indio Muerto, and on the south by a range of hills which extends to the Cerro Negro and Carrizalillo. It consists of narrow valleys, and communicates with the sea in front of the Pan de Azucar.

The fourth basin comprises the dry channel of the Salado river, and the undulating tracts of land which formerly discharged their waters into it.

The range which branches from the Cerro del Azufre

skirting the elevated plain of Tres Puntas and joining the coast-chain close to Las Animas, constitutes its southern limit. The surface of this basin is very undulating, with no extensive plains, but consists of a series of long and narrow valleys.

Although the four ranges which form the boundaries of these basins attain very considerable altitudes, they nowhere assume, in the interior of the desert, the character of precipitous hills, but rather that of hills of rounded form and gentle slopes. From these jut numerous spurs, the general direction of which are north to south, and north north-east to south south-west.

These spurs subdivide the large basins into plains of lesser magnitude, some of which are enclosed on all sides, and have apparently been the sites of ancient and extensive lakes.

Such is the general description of the "Desert."

The plains are more elevated the nearer they approach the chain of the Andes. At a distance of 100 kilometres (62 miles) from the coast their height surpasses that of the maritime range, the rise of the land is gradual and almost uniform. The plain of Cachiyuyal, distant about 44 miles from the sea, is 4500 feet above its level, whilst that of Cachinal de la Sierra, 100 kilometres from the coast, attains an elevation of 7400 feet. The first rise corresponds with an average slope of 2·28 per cent., and the last of 2·27 per cent. or about 1 in 44.

This disposition of the land is very favourable for the construction of railways, the more so as the soil is, for the most part, alluvial. Hence the cost of earthworks will be comparatively small.

II. Geology.

The geological structure of the desert of Atacama is one of marked uniformity. The different formations are deposited in parallel beds dipping approximately north and south, and in such wise that in every part of the desert the same formations are found succeeding one another in the same order.

In the neighbourhood of the sea, and forming the western watershed of the mountain-chain, which runs parallel with the coast, there occur stratified rocks which all belong to the azoic and palæozoic periods; these are gneiss, schists, silicious rocks, grauwacke, and coloured sandstones. These stratified rocks are always much inclined, and are frequently intersected by masses of interposed plutonic rocks, amongst which may be noted syenite and labradorites.

These last are especially met with near the sea, where they constitute the greater part of the reefs, as also some small islands.

Wherever the stratified rocks have come into contact with these plutonic masses they have been very much altered in structure and composition; they almost always assume the aspect of porphyry, and are traversed by numerous veins of quartz and "espidota." Hence it is difficult to recognize them at first sight, since the primitive character of their stratification has almost entirely disappeared.

At a distance from the coast, varying from eight to nine miles, the plutonic rocks begin to predominate; passing away from the narrow margin occupied by the stratified formations, they extend to the foot of the Andes, occupying nearly the entire area of the central depression.

Where the stratified formations come in contact with the hills formed by the plutonic rocks the latter present a peculiar aspect. They are intersected by numerous lines of a dark colour, which preserve a certain parallelism, and which have gained for some of them the denomination of Cerro Votado. These lines are simply interleaved beds of the stratified formations which have been split up and become involved in the plutonic mass.

The nature and age of these plutonic masses are not the same over the whole extent of the desert. The rocks succeed one another in a definite order, the least ancient being situated more to the eastward, whilst on the eastern slope of the coast-chain the syenitic rocks crop up, constituting the axis of this range of mountains. They are less observable in the great central depression, and are by degrees replaced by augitic porphyries and amygdaloid rocks, the latter being replaced in their turn by trachytic rocks. Lastly modern trachytic rocks, viz., pumice and lavas, are found in the highest parts of the Andes chain. Such is the general disposition of the plutonic rocks; but there are some here and there, even on the edge of the sea, such as syenites and augitic porphyries.

Close to the port of Chañaral syenites may be seen to intersect the stratified rocks, in some places forming considerable masses, in others contracting into narrow dykes, which penetrate between the strata as offshoots from the principal masses. In Taltal it is the augitic porphyries which intersect the strata and have transformed them at the points of contact into amygdaloid rocks.

The stratified formations again make their appearance near the base of the Andes, but differ from those which are found in the immediate neighbourhood of the coast; they belong to an epoch less remote, and lie almost always above the red sandstone. Some reddish clays, referrible to the Triassic formation are the first found, resting immediately upon the red sandstone and beyond the region occupied by the plutonic rocks. These beds occasionally approach the coast-chain, and may be seen near the Papaso mines at the nitrate-works of Cachinal and Aguas Blancas.

Further to the eastward, and extending over the slopes of the Andes, appear the calcareous formations of the "Jurassic" period. Throughout the whole of this region the stratified rocks have experienced numerous upheavals, they are broken up by the plutonic rocks, and do not appear except in belts of greater or less width, extending in the direction of the transverse ranges, and approaching more or less to the coast-chain, as occurs with the calcareous formation of Tres Puntas, that of Florida, which is still nearer to the sea, those of the Encantada and of Sandon, more to the eastward, and finally those of Cachinal de la Sierra and de la Palestina.

In all these belts the calcareous formation presents the same characteristics as in the immediate neighbourhood of Copiapo and of Tres Puntas.

The interior portions are composed of calcareous sandstone and of jaspers, with superimposed beds of marl and limestone with their numerous fossils.

Such is the order and relation of the ancient formations of the desert. Near the coast the stratified palæozoic rocks; in the intermediate part the plutonic rocks; the Jurassic formation upon the western slope of the Andes; and in the highest part of this vast aggregation of mountains, are the volcanic formations in the midst of which rise up the extinct volcanos of Azufre, Doña Ines, Chaco, and Llullaillaco. But there is still another formation, much more modern, which imparts to the desert its characteristic aspects. What especially attracts attention on visiting the central region for the first time is its nakedness and uniformity, the plains and hills being covered with sand or small loose stones. On examining these small stones carefully, it is seen that they all retain their angular shape, and that they could not have been derived from alluvial deposits like those which cover the plains of southern Chile.

Moreover, rising at intervals, there are large rocks of irregular shape like the ruins of ancient buildings with their windows and high and slender spires, forming a marked contrast to the undulating and continuous outline of the hills.

This vast amount of débris and scattered rocks has been the result of disintegration of the plutonic rocks. Whilst the hills became worn down by slow degrees, the more resisting portions remained standing out above the surface as isolated masses of rock. Their destruction has been mainly effected by the constant changes of temperature. The plutonic rocks, subjected during the day to the powerful action of the sun, are heated as much as 120° F., and at night are cooled rapidly to 38° F. in summer and below freezing-point in winter. Thus subjected to a series of repeated expansions and contractions, disintegration took place. The rocks on the summits of the hills split into leaves which open like those of a book; others split off in concentric layers which separate from the mass and fall in the form of small scales covering the slopes of the hills. Finally, by the action of the atmosphere on the felspathic rocks, the felspar and all the separated scales are reduced to small dust.

One only of the constituents of these rocks resists destruction. This is silica, in the form of quartz or chalcedony. At the places where the amygdaloids occur (and these always accompany the porphyries and the trachytes) the chalcedony which they contain remains on the surface. Hence the origin of the fragments which are found so profusely strewn over it that sometimes extensive plains are covered with them. Finally, the infrequent but heavy rainfalls carry down to the plains all this loose detritus, which forms in course of time the thick layers which cover them.

To the same causes must be attributed the rounded forms and the gentle slopes of the hills, their salient points having been those which were most exposed to the action of the sun.

From this great formation of detritus other deposits are derived which, in sight of their importance, deserve a closer study.

III. Deposits of Nitrate of Soda.

The nitrate deposits are found dispersed in the central part of the desert from 26°·30 to nearly 24°, southern latitude. They are situated in the upper portions of the plains which lead into the great basins or into the vast enclosed tracts of level ground which appear to have formed the sites of ancient and extensive lakes. They only occur at a certain distance from the channels of the ancient rivers, and whether in the valleys or in the plains the richest portions are not met with in the centre but on the narrow belts of the surrounding rising ground.

This disposition appears to be due to the great solubility of the Nitrate of Soda, and in the lowest parts where the water has been detained its quantity appears to have been sufficient to dissolve the salts and carry them below the surface by filtration.

The strata in which the Nitrate of Soda exist are recognizable by certain characteristics, and present themselves under two totally different aspects. Those which are known by the name of "Salares" are recognizable at a great distance by the quantity of common salt. This occurs in rounded masses rising above the general level of the surface; these are full of cavities containing common salt, sulphates of soda and of lime, combined with a sensible quantity of earthy matter.

It is under this bed of salt that the Nitrate of Soda is met with, disposed usually in layers, the thickness of which varies from 10 to 50 centimetres. Its colour is dark, its structure porous, and it always contains earthy matter. The central portion of these deposits is poor in saltpetre, and it is only on their borders that this is found in greater purity and larger quantity.

In the other class of localities the saltpetre does not appear on the surface, which consists of a bed of earth and small stones, but there are two certain signs by which its presence underneath is indicated. The first is the existence of small natural pits, which occur at intervals over the surface.

These are especially met with in depressions of the ground where the waters have been detained, and been thus enabled to dissolve the Nitrate; consequent upon evaporation the soil sank and cavities were formed. The second sign consists in the occurrence of numerous surface-fissures crossing in all directions, and dividing the surface into an infinite number of small polygons, which have imparted to it a very strange appearance by reason of the curious designs produced by stones collecting in the fissure. This fissured formation is the result of the crystallization of the mass of saltpetre, causing shrinkage and a separation into wide prisms, the lines of division of which have extended to the surface.

The presence of chalcedony has also been regarded as an indication of the existence of saltpetre, and it is certain that this is met with in large quantity in some saltpetre-refineries; but there are also many cases in which it is not found, and if what has been said previously in reference to the origin of the chalcedony is borne in mind, it will be seen that the presence of this mineral has no connexion with that of the Nitrate.

Below the surface-soil, which varies from 1 to 5 or 6 decimetres in thickness, is found a light-coloured bed of plaster (gypsum?), including numerous small stones, known to the saltpetre-hunters as *costias* or crusts; this is from 2 to 4 decimetres thick, and forms the immediate cover of the saltpetre deposit, which is extremely variable, both in quality and thickness, the latter ranging from 1 or 2 decimetres to 2 metres and upwards.

In the same deposit patches of ground are found in which the saltpetre is in a compact form, and simply mixed with salt and sulphate of soda, while in the others it is more or less mixed with earth. Below a similar class of deposit to that overlying it is occasionally found, while in other cases it lies directly on the bed-rock of the neighbouring hill. In the latter case the nitre is usually found to be of a better quality than when it is deposited on a substratum of plaster.

Of the saltpetre deposits known up to the present time that adjacent to the lagoon of Cachiyuyal is the nearest to the sea. It extends from a point 6 kilometres S.E. of the lake, and extends to the Cerro del Hornillo, and is situated upon a low hill rising on the west side of the bed of the old river of Cachiyuyal, forming a narrow belt of 60 or 70 metres wide, for a length of about 8 kilometres. The maximum thickness of about 1 metre of nitre is found in the central portion, a gradual thinning away being observed in either direction. The mineral is of a yellowish colour, contains a great deal of salt, and does not assay above 28 per cent.

Following a north-easterly course for a distance of 26 kilometres from the lagoon of Cachiyuyal the saltpetre works of Gonzalez are reached. These are situated in a small depression of the ground, between the hills rising to the north of the road from Cachinal de la Sierra. This is small in extent and irregular in character, but yields an excellent quality of nitre, although somewhat mixed with sand. The deposit lies upon porphyritic rock, in which it forms a series of veins.

The saltpetre works of Baron are about 12 kilometres north-east of the preceding, being separated from them by an elevated plain covered with agates. This has been but slightly explored, the section exposed in the basin of a little more than a metre in thickness showing the nitre resting upon porphyritic rock, and penetrating it to a certain depth. The mineral is mixed with 20-35 per cent. of sand, and yields a pure white product, more than 45 per cent. being soluble.

The deposits that have been most completely explored are those of the Callega Company, Gunan, situated about 16 kilometres north of Baron, in an extensive plain, bounded by hills and mountains, the centre of which appears to be the bed of an old lake. This portion is occupied by a thin bed of nitre, which, however, increases in volume as it approaches the high ground. On the north side numerous trial-pits have been sunk, exposing sections of the nitre-bed varying from 1 to 2·60 metres in thickness. The mineral is compact and yellowish, containing notable quantities of salt and sulphate of soda, the average of the assays for nitre being from 23-30 per cent. The substratum is rather hard gypseous conglomerate. On the northern side the conditions are somewhat similar; but the thickness of the bed is less rarely exceeding one metre. In the portions that have been thoroughly explored the underlying rock varies, sometimes being the gypsum-bed and sometimes a decomposed porphyry, the latter giving the highest quality. Beyond the hills bounding these deposits to the north another place is found in which a new deposit of nitre has been recently discovered; but little has, however, been done as yet in the way of exploration trials, and those only on the eastern portion. Both as regards position and quality, there is considerable analogy between it and that of Baron, the nitre, though mixed with sand, being of a superior quality, assaying over 40 per cent., with but little salt and sulphate of soda.

Lastly, two other deposits have been found to the eastward of the last mentioned, in the direction of Cachinal de la Sierra. The first, known as the Descubidora de Buñado, 24 kilometres from the lagoon, is situated in a plain about 12 kilometres long, and 3 or 4 broad, which shows a bed of saltpetre at several points of about one metre in thickness. The quality of the mineral is fairly good, as, although a little red, it yields 36 to 42 per cent. of nitrate. The second, known as the first saltpetre Works of the Guzman Company, is of considerably less importance than the preceding; it occupies a small extent of ground between two hills to the south of La Descubidora, and from the small amount of exploration hitherto carried out, it is difficult to form any exact estimate of its value. In a few of the trial-pits nitrate has been found; but in others the mineral is only sulphate of soda.

Such, then, are the general characters of the nitre-works of Cachinal.

The great cost and difficulty of making more detailed examination has hitherto prevented their being systematically carried out in the manner that the importance of the subject requires. In most cases the explorers have been contented with demonstrating the presence of nitrate without showing either the thickness of the bed or the quality of the mineral. The former point would, on account of the great irregularity of the deposits, have required a great number of pits to be opened in each district, if anything like an exact estimate as to quantity was required. Two principal results appear, however, to be established from the discoveries already made, namely—that the greatest mass of mineral is to be looked for not in the central portion of the plains, but on the slopes of the hills; and bearing this in mind it is evident that in many instances the points selected for examination have not been judiciously chosen, and that the sites of probably more important deposits have been passed aside. In the second place it is proved that these deposits are of considerable extent, being first met with at the eastern base of the coast-chain, where they extend across the supervening plains to the foot of the Andes.

IV. Aguas Blancas.

In lat. 24° 6′ S., and at a distance of 65 kilometres from the coast an extensive saliferous tract of country is met with, which extends southwards to the marshes of Aguas Blancas, and eastward to the point known as Cuevitas or Agua Dulce, covering an area measuring 38 kilometres from east to west, and 16 kilometres from north to south. Below the superficial crust of salt Nitrate of Soda is found in their beds of 1 to 2 decimetres thick, mixed with soil, salt, and sulphate of soda. Towards the edge of the deposit the mineral thickens to 5 or 6 decimetres, becoming more compact and of better quality, in some instances assaying up to 23 per cent. Very little is, however, known of this district up to the present time. As a general result it has been observed that the quality of the mineral improves southwards in the direction of the marshes, and also south-eastward towards the Cordillera of Varas, where probably such deposits may be expected, it being a well-ascertained fact that nitrate is not found, or at any rate not of good quality, in the middle of salt-covered plains. Future exploration may therefore be most usefully carried out in the western, southern, and eastern portions of the district in question.

V. On the Origin of Nitrates.

It will not be unnecessary to consider this point somewhat in detail, as the conclusions arrived at, though necessarily speculative, may be of value in facilitating future discoveries. The constant association of common salt with nitrate seems at first sight to point to a probable marine origin of those minerals; but on going further into the question it becomes evident that the total absence of limestones and other stratified rock, as well as of marine shells such as might have been expected in deposits presumably formed in a shallow sea-bed, is incompatible with such a mode of formation. Furthermore, the Nitrate is, in many places, found associated with beds of pebbles, which could not have been expected in a deposit formed either by slow sedimentary action in water or by the concentration of a saline solution by evaporation. The mineral does not occupy the lower tracts of the district, but is generally accumulated on the hill sides, and is even found at great altitudes, as, for instance, at the mines of Paposa and on the Cordillera of Maricunga, the latter being more than 13,000 feet in altitude. It is evident therefore that it is to local conditions that the production of the nitrate and its accompanying saline minerals are due, and of these the most important elements are furnished by the felspathic constituents of the surrounding rocks, which, by their denudation, furnish the felspathic sand forming the slopes of the nitrate plains. The felspathic minerals contained in these rocks are labradorite, albite, and oligoclase, the first containing a considerable quantity of lime, the second from 8 to 10 per cent. of soda, and the last both soda and potash, and therefore furnishing bases for all the salts found in the deposits. Sulphuric acid for the formation of sulphates is probably furnished by the oxidation of pyrites, an invariable constituent of these rocks; and chlorine is constantly present in volcanic emanations, the waters derived from the trachytic areas containing large quantities of soluble chlorides. The formation of nitric acid, although not so readily explicable, is, according to the results of the experiments of Cloes, to be accounted for by the property possessed by alkaline carbonates of transforming atmospheric nitrogen into nitric acid in the presence of other oxidizable matters. It is also well known that under the influence of the atmosphere felspars become changed into china clay with the loss of those alkalies which are converted into carbonates, any protoxide silicates of iron existing in the other constituent minerals of the rock, augite, hornblende, &c., turning at the same time by oxidation into peroxide compounds, which conditions are precisely those required by the theory of nitrification in question. If we remember what has been previously said as to the rapidity of decomposition of the rocks in the desert area, the production of nitrate, and its position at the foot of the hills will be easily understood.

The rocky hills, by their gradual disintegration, are reduced to a coarse sand, forming a talus on the slope, which by the rare but heavy rains of the district is removed to the plains. These felspathic sands, by decomposition, give rise to a soil consisting of china clay, brown iron-ore, gypsum, common salt, and carbonate of soda, the latter being changed in its turn into nitrate, which in subsequent rainfalls is dissolved and infiltrated into the ground up to the rise of the hills, while the gypsum, being considerably less soluble, remains, chiefly in admixture with china clay, forming the crust or cover of the nitre-beds, being found not only in the plains, but up to the highest points of the hills; for whenever the surface-sand is removed at any point a white porous substance consisting entirely of gypsum is invariably found. The infiltrate solutions during the long rainless intervals being slowly evaporated, deposit their salt in a crystalline form on the soil near the surface, forming a more or less intimate mixture with the original absorbent sand and earthy matters.

It has thus been shown that the nitrate is entirely due to the decomposition of felspathic rocks, and as these form the central portion of the desert of the Rio Salado, on the 24th parallel, it cannot be doubted that in addition to those deposits already known, others must exist in considerable numbers south of the 26th and north of the 28th degree of south latitude, and that such deposits may be easily discovered by explorers guided by the indications given above.

VI. Probable Quantity of Nitrate.

The trial workings at Cachinal and Aguas Blancas are so very small in extent as compared with the Nitrate-area that it is quite impossible to form even an approximate estimate of the possible quantity of nitrate at these points.

All that can be said is that it is certainly very large; and to give some idea upon this point I should confine myself to the examination of 2nd and 3rd works of the Guzman Company, which have been the best explored. These include 600 hectares, or 6 millions of square mètres of ground, within which several pits sunk at points not specially selected show thicknesses varying from 1 to 2·5 metres of the Nitrate-bed. Taking the smaller of these figures we obtain a volume of six millions of cubic metres, or in weight (the specific gravity being about 2), 240 millions of quintals of crude Nitrate, which, at supposed yield of 20 per cent., corresponds to 48 million quintals of refined Nitrate of Soda. Taking a million of quintals as a fair yield from these places annually, it is evident that these two concessions would be workable for forty-eight years before exhaustion. It will therefore be seen without the slightest fear of exaggeration that the Chilian portion of the Desert of Atacama is capable of producing very large quantities of nitrate for more than a century.

VII. Methods of Working.

Although the method to be followed in the working of any given deposit, in order to utilize it to the greatest advantage, must, as a rule, be determined by the skill and practical knowledge of the parties immediately concerned, it may be useful to make a few general remarks upon this subject.

The Nitrate-deposits are not such as can be compared to those of coal or other stratified minerals, being essentially irregular, both as regards quality and quantity; and as in all other industrial enterprises, one of the most important points is the reduction of first outlay to a minimum, the working should always be commenced at those points where the best quality of mineral is found. This requires a preliminary examination of the ground, which, however, should be comparatively inexpensive, the nitrate being found at shallow depths, under a covering not usually of any great hardness.

The methods usually followed in determining the value of crude Nitrate being both complicated and inexact, I have devised the following improved plan, which is more simple, sufficiently accurate, and suited to the requirements of the persons interested in this particular industry.

The sample of mineral is mixed with an equal weight of charcoal, and deflagrated at a dull red heat in a crucible, when the nitrates are converted into carbonates, and may be dissolved out in water, the residue is thrown upon a filter and washed until no further alkaline reaction is apparent in the waste water. The proportion of carbonates in the liquid is then determined by the ordinary methods of alcalimetry, 1000 parts of carbonate being the equivalent of 1602 of nitrate. This method is so simple that it may be used on the ground. A more approximate method is founded upon the phenomena observed during the reaction. When less than 15 per cent. of nitrate the decomposition takes place without visible deflagration; between 15 to 25 per cent. the deflagration is apparent, but very moderate in character; from 25 to 40 per cent. it is more vivid, and at higher percentages than 40 the action is so violent that a portion of the material in the crucible may be projected unless care be taken to moderate it by increasing the amount of carbon to double that of nitrate. In any case care must be taken to keep the temperature as low as possible, otherwise a portion of the sulphate of soda may be reduced by carbon to sulphide, which passing into solution would show too high a result in the subsequent alcalimetrical determination.

The working of Nitrate-deposits is based primarily upon the increased solubility of Nitrate of Soda in water with increasing temperature. At $10°$ C. water dissolves 78 per cent. by weight of nitrate, and at $100°$ C. 177 per cent., the difference, or 99 per cent., separating in crystals when the solution has cooled down to the lower temperature. Fuel is therefore of prime necessity, and forms the principal item of expenditure in the manufacture. Possibly solar heat might be substituted in some way, as it would not be difficult to raise the temperature of water by the use of specially constructed apparatus, during the hotter parts of the day, to $60°$ C., when it would dissolve 131 per cent. of nitre, depositing 53 per cent. when cooled to the lower temperatures prevailing at night.

The richest mineral is not always the most advantageous, as from its compact character it is not readily soluble and requires to be broken by mechanical means into small pieces before treatment. The presence of sand, which loosens the mass, and allows access to the dissolving water, is therefore an advantage.

It may be considered probable that in some deposits the necessary amount of water for working will not be forthcoming on the spot. In such cases it will be necessary to establish the nitre-works upon the nearest available watercourse. Care must, however, be taken not to allow them to approach too near, or in the upper portions of such water-supply, in order to prevent the residues from infiltrating and fouling the fresh water to such an extent as to render it unfit for domestic use.

In addition to Nitrate of Soda certain necessary products are annually obtained in the nitrate-works, which may be utilized considerably, such as sulphate of soda, which may be employed in making soda-ash and iodine, the latter being met met with at times in such quantity as to be susceptible of profitable extraction.

VIII. Means of Transport.

The successful working of any branch of industry where the article produced is low in price, is in great part dependent upon the cost of transport. It is therefore of prime necessity in the case of the nitre-works in question to establish channels of communication that will allow this condition to be realized. From this point of view the Nitrate-districts, and those of the north more particularly, are but slightly favoured. Between the 26th and 24th parallels of south latitude the coast-line of mountains has a mean elevation of 1200 metres, and throughout this length there are only three transverse valleys through which lines of communication can probably be established with the interior. The first of these, lying to the north of the valley of Remiendos, which has already been examined by a special commission, would be available for the construction of a waggon-road, but is too rough for railway purposes.

This valley, at its mouth on the coast, in lat. 24° 21′ S., bears in a north-east direction (looking up) for a short distance, then turns between east and south east until it reaches the first series of plains 28 to 30 kilometres distant. Further on, as far as the pass of the Cardones, the ground is without special difficulties; but the rise to the pass is rather a long one, and there is about an equal descent to be made on the other side. The distance from the entry of the plain to the pass is 46 kilometres, and thence to the marsh 22 more, so that the length of road for vehicles to the coast would be 96 kilometres, and on account of the unfavourable nature of the gradients 4 cwt. would be a full load for a cart.

The valley of Paposo, on the 25th parallel, presents still greater difficulties, and the cost of a good waggon-road on this line would be very great.

The third valley, that of Taltal, in lat. 25° 24' S., is deeper than the preceding, and cuts right through the coast-range, so that the ground has a general slope from the sea to the foot of the Andes. The height of the lagoon at Cachiyuyal, situated at the entry of the plain, is 1371 metres above the sea, and its distance from the coast 60 kilometres, which gives an average slope of about 1 in 47.

The soil of this valley is essentially gravelly, and the bottom slopes uniformly, with the exception of a short distance above Las Breas, where there is a somewhat difficult piece of ground, which could, however, be overcome without much trouble or expense. This section of the low valley is perfectly well suited for railway construction; and further up, between the lagoon of Cachiyuyal to the nitre-beds of Cachinal, the slope is much lower, the distance and difference of altitudes between the former point and the second pampa of the Guzman Company being 40 kilometres and 592 kilometres respectively, corresponding to a slope of about 1 in 66.

From the Nitrate-works to the sea the slope is nearly uniform, and sufficient to allow loaded waggons to descend by gravitation, so that engine-power would only be necessary to bring back the empty waggons, and the small upward traffic in provisions, &c., to the mines and works. The valley of Taltal is therefore the only one fit for the construction of a railway on the scale required to satisfy the demands of a large mineral industry. A short branch might be made from the lagoon to the nitre-works of Bañados and Olivos; but the main line must follow a northerly direction, as being that of the important deposits; and if, as there is every reason to suppose, others should be found still further north, the line might be continued as far as those of Aguas Blancas.

The port of Taltal seems therefore likely to become the principal point for the export of the nitrate and other products of the interior of the desert, and from this point of view it merits the special attention of the Government. It is above all necessary to be careful in granting concessions for the railway, so as to limit the maximum rates of transport in order to protect the producer of nitrate against possible overcharge by the owners of the railway. The maximum rate per quintal should not exceed two centavos (10 centimes) per 10 kilometres, as a higher rate would so trench upon profits as to prevent capitalists from embarking in the manufacture of nitrate.

IX. Deposits of Guano.

In addition to the nitre-beds the desert contains deposits

of guano of a certain extent. There having been no particular inducement to private individuals to make discoveries of these, we are without much acquaintance with any but those in the Nitre-district belonging to the Guzman Company. Here they form two bands, whose breadth and thickness have not been determined, with a linear extension of about 1 kilometre. The guano differs from that of other places, especially by its richness in organic matter. A sample collected by me contains 28 per cent. of organic matter, chiefly borate of ammonia, as well as 9 per cent. of nitrate and phosphate of ammonia, a little nitrate of soda, and 8 per cent. of sulphate of lime. Or, according to the analysis, it contains 42 per cent. of fertilizing matter, and these of the most energetic character, being chiefly ammonia. The remaining 58 per cent. consists of sand, with some hydrated peroxide of iron. The guano-beds of the desert are therefore well worthy of attention from their specially good qualities, or whenever a demand shall arise for their utilization, numerous other discoveries will doubtless be made. The most efficacious means of encouraging the research for such deposits would be to declare them free to work, and to levy an export duty upon the produce, rather than to continue the present practice of farming the monopoly to particular individuals. The presence of guano is indicated by certain superficial characters, in like manner to that of nitrate. It is almost always found near the nitre-beds, and where it exists the soil is usually yellow, as though it contained a considerable amount of peroxide of iron. Furthermore, these deposits are so close to the surface that they are often laid bare by the horse's hoofs in riding over the ground.

X. Borate of Soda and Lime.

Another product which will in time become of considerable importance, is the borate of soda and lime (ulexite) which has been discovered at different points in the desert. Up to the present time only two deposits are known, that of the lagoon of Maricunga and another near the lagoon of La Ola. But as it is well known that boracic acid is a common product of volcanic activity, it may be supposed that others will be found in the neighbourhood of other volcanoes of the Cordillera, as, for example, near those of Chaco and Llallaillaco.

XI. Deposits of Metals.

The minerals previously considered occupy a large extent of the desert area; but in addition to these there are others of a less value, namely, metallic minerals, which are profusely distributed over the ground, there being scarcely any portion of the coast-range, from the 27th parallel to to the Chilian frontier, which does not contain some mineral veins; in proof of which it will be sufficient to mention the names of the mines of Salado, Las Animas, Cerro Negro, Carrizalillo, Cachiynyal, Paposo, and del Cobre, the most characteristically abundant amongst the minerals being copper-ores.

These are found more particularly in the eastern slope, where the syenitic rocks are traversed by intrusive bands of labradorite and augite porphyries, generally in veins between the two classes of rock, as though the cupreous matters had filled the fractures produced by the later igneous intrusions. In other places they penetrate and are included in, the porphyries, often forming considerable masses of ore, as at Carrizallilo. A very marked difference is observed between the minerals found in the veins, according as they are in contact with labradorite or augite porphyry. In the former case, after passing through the oxides, oxychlorides, and silicate of the Gozzan, the ore is chiefly copper-pyrites, while in the latter variegated copper ore, grey sulphides, and ruby-ore prevail, the whole being, as a rule, markedly argentiferous.

The deposits of copper ores are not confined to the coast-range, but are also found at the foot of the Andes, where, from the presence of augitic porphyries, the minerals are of the richer class of sulphides mentioned above. The distance of the mines from the coast, and the consequent high price of fuel, has hitherto prevented their being worked to profit ; and only a single one, that of Sandon, is systematically worked. It is also in the western slopes of the Andes, and more particularly in rocks of Jurassic age, that silver ores are principally found, the veins bearing these minerals being intimately connected with the limestones of that period, and the subsequent eruptive masses of trachytic and augitic porphyry. In most instances the veins are also at the contact of the two classes of rock, as in the former cases.

If we recall what has been previously said concerning the development of the Jurassic series in the desert it will be easy to understand the position occupied by these argentiferous veins. We have seen that this formation, besides appearing on the western slopes of the Andes, extends southwards in the transverse ridge, which sometimes approaches very closely to the coast-chain. The well-known mines of Chimbero and Tres Puntas, are examples of deposits situate in these transverse ridges. Further north are the mines of La Florida, situated in a band of limestone, forming part of the range that closes the basin of Salado to the north. A bed of augitic porphyry, running east and west, has uplifted two series of limestone beds, and in its neighbourhood the silver veins are found.

Nearer to the base of the Andes we come to the mines of Sandon, and lastly, near the Chilian frontier, in about $24°$ S. lat., are the mines of La Palestina.

In addition to the above, numerous veins of argentiferous galena are known, but up to the present time they have scarcely received any attention, on account of the difficulty of working them to a profit. It may be hoped that the development of lines of communication consequent upon the opening-up of the nitrate-district, will facilitate the means of exploring the other minerals of this region, so that they may, in their turn, be actively worked.

XII. Conclusion.

The desert of Atacama presents a great field for the development of mining industry, and from this point of view merits the serious attention of the Government. Although it is true that private individuals should be left full liberty of action in the way of exploring and opening works, the natural obstacles of the district are such as to render such operations costly and difficult, and therefore the action of the Government will be required to effect the necessary improvements. I have already mentioned the port of Taltal as the most important point on the coast, and the true gate of the desert; but there is no means at present available for reaching this place. The steamers of the Pacific line do not call there, and miners wishing to reach Cachinal and other places in the interior, have to make a long land journey from Chañaral, through a country without resources; whereas if the steamers called at Taltal the journey would be shortened by two days, and the increased facilities thereby afforded of procuring the necessaries of life would speedily attract a numerous population to the spot.

A. PISSIS.

Santiago, June 28, 1877.

ASSAYS AND ANALYSES OF SAMPLES OF SALTPETRE REFERRED TO IN THE PRECEDING REPORT.

Nitrate from Aguas Blancas (pale yellow colour).

Insoluble in water (clay and gypsum)	9 per cent.
Soluble in water	91 ,,

Containing—

Nitrate of Soda	15	per cent.
Sulphate of Soda	56	,,
Chloride of Sodium	24	,,
Sulphate of Alumina	3	,,

Nitrate from the Pampa de Lavanderos, Cachinal.

Insoluble in water (felspathic sand, quartz, clay, and gypsum)	51 per cent.
Soluble Salts	49 ,,

Containing—

Nitrate of Soda	76	per cent.
Sulphate of Soda	24	,,
Chloride of Sodium	6	,,
Sulphate of Alumina	4	,,

Nitrate from Baron.

Insoluble in water (sand, clay, quartz, and gypsum) .... 47 per cent.
Soluble salts .... 53 ,,
Containing—
    Nitrate of Soda .... 52 per cent.
    Sulphate of Soda .... 6 ,,
    Chloride of Sodium ...... 34 ,,
    Sulphate of Alumina ...... 5 ,,
    Iodine .... traces.

Nitrate from the Second Pampa of the Guzman Company.
    Insoluble substances ...... 53 per cent.
    Nitrate of Soda .... 12 ,,
    Sulphate of Soda .... 9 ,,
    Sulphate of Lime .... 9 ,,
    Sulphate of Alumina ...... 8 ,,
    Chloride of Sodium ...... 6 ,,
    Water .... 1 ,,
    Magnesia .... trace.

Nitrate from the Concession Peña, Aguas Blancas.
    Insoluble substances ...... 9 per cent.
    Nitrate of Soda .... 5 ,,
    Sulphate of Soda .... 68 ,,
    Sulphate of Alumina ...... 6 ,,
    Chloride of Sodium ...... 9 ,,
    Water .... 1 ,,

Nitrate from the Pampa of Lavanderos.
    Insoluble substances ...... 6½ per cent.
    Nitrate of Soda .... 17 ,,
    Sulphate of Soda .... 6 ,,
    Sulphate of Alumina ...... 2 ,,
    Sulphate of Magnesia ... 3 ,,
    Chloride of Sodium ...... 6 ,,

Nitrate from the Second Pampa of Cachinal.

Insoluble substances......	23 per cent.
Nitrate of Soda............	32 ,,
Sulphate of Soda	10 ,,
Sulphate of Lime.........	6 ,,
Chloride of Sodium	22 ,,
Water........................	2 ,,

ASSAYS OF CERTAIN NITRATES.
(The refined salt expressed in percentages.)

Nitrate of Aguas Blancas...................	13 per cent.	
,, Pampa de Lavanderos.........	37	,,
,, Baron	27	,,
,, La Descubridora (Bañados)	42	,,
,, Cachiyuyal	6	,,
,, the Third Pampa...............	13	,,
,, Baron	30	,,
,, Aguilar...........................	15	,,
,, Aguas Blancas (Peña).........	6	,,

Letter from the Minister acknowledging the receipt of the preceding Report.

Santiago, 6th July 1877.

"I have read with much interest the Report submitted by you containing the results of the mission with which you have been entrusted by the Minister of the Interior, for the purpose of examining the nitre-beds recently discovered in the portion of the Desert of Atacama included within the territory of the Republic, and I am pleased to recognize that you have amply justified the confidence which the Government, relying upon your experience and wide knowledge, have felt in placing in your hands the study of the important effect the natural riches of the territory may exercise on the well-being of the national industry. The fatiguing and delicate mission which you have carried out at the cost of great personal exertion has confirmed scientifically the information upon which the action of the Ministry was based, with the ultimate view of opening new fields for our industry in a desert region both uninhabited and unnoticed by capitalists; and I am persuaded that this region will soon be the seat of an industrial development that will increase the national wealth and put an end to the period of difficulties which we are now passing through. I pray you to accept my congratulations on the manner in which you have accomplished your mission.

JOSÉ VICTORINO LASTARRIA.

No. 2.

Report presented by the Engineer, Don JOSÉ R. MARTINEZ, *to the Council of Mines in the Session of June* 14, 1877.

I have the honour to lay before the honorable Council of Mines the following Report of the mission with which I have been entrusted, as a member of the commission for exploring the nitre-beds of the Chilian desert on the coast-range of mountains and to the south of the 24th parallel.

The first locality visited was Carrizalillo, in which neighbourhood we were informed of the existence of a deposit of Nitrate. At a short distance to the north of this place is a broad flat valley, the end of which is covered with white stratified deposits resembling, when seen from a distance, nitrous efflorescence. These, upon closer inspection, proved, however, to be only beds of gypsum of great thickness, in places covering an area of several square kilometres.

On the return from Carrizalillo, a visit was paid at the desire of the chief of the expedition, M. Pissis, to the mines of La Florida for the purpose of investigating the character of the deposit. According to his opinion these mines are contained in limestone strata of Jurassic age, which have been uplifted by an outburst of labradorite, the veins being most productive at the line of contact with the latter rock.

From Chañaral we proceeded by sea to Taltal, where, before starting for the interior for the purpose of investigating the nitre-beds of Cachiyuyal and Cachinal de la Sierra, one of our party, M. Villeneuve, laid out the plan of the settlement which will be required in this place, should the development of the Nitrate-works proceed in the manner that we are justified in expecting them to do.

The port of Taltal is an excellent one, the water being perfectly smooth and with sufficient extent of beach for the requirements of any extent of population that may be reasonably required for the conduct of the trade.

The first traces of Nitrate are found at a place called the Pique de Cachiyuyal, where it is found in small quantity mixed with common salt. It is also found in reniform concretions, but of small thickness, in a valley of considerable extent running from south-east to north-west. Below the saline deposit is an impermeable clay-bed, whose presence explains the apparently anomalous phenomenon of the existence, at a small depth, of fresh water below a surface covered with salt.

The deposits of nitrate extend from Cachiyuyal to Cachinal de la Sierra, occupying the bottoms and sides of a series of valleys of gradually increasing altitude as the distance from the coast increases, the highest points being to the eastward. The first of any great extent is to the north-east of Cachiyuyal, on the side of the montain of Hornito; this, however, has been but very slightly explored, and the mineral is very full of common salt.

Other deposits are found to the east and north-east of Cachiyuyal, and in that part which has lately been named Cachinal de la Sierra. The principal concessions are known as the first, second, and third Pampas of Messrs. Guzman, Barazarte and Company, and that of Messrs. Oliva and Gonzalez. The first pampa lies to the east of the valley of Cachiyuyal, and about seven or eight leagues from the Pique; the second is to the north-east of the same valley, in a plain of some extent, where a bed of Nitrate, averaging one metre in thickness, is found to extend continuously for a length of 4 or 5 kilometres, and a breadth of 1 kilometre. In the third, situated to the north of the preceding, Guano is found which, judging from the sample obtained, is of a very superior quality.

This Guano was first obtained by M. Callejas, a partner of the concessionaires; but the actual deposit has not yet been discovered, so that nothing can be said as to the extent or thickness of the bed. The only indications afforded are yellow bands, from 4 to 6 metres broad, appearing at intervals upon the ground. In any case the presence of this Guano allows us to infer the existence of a considerable quantity in the neighbourhood; and that it has not yet been found is to be accounted for by the very small amount of exploration actually accomplished in the district.

The concession of Messrs. Oliva and Gonzalez lies to south of the first pampa. Here very pure crystallized Nitrate is found in a bed of about a metre in thickness, which has been but very slightly explored, although the valley is of very considerable extent; and all the circumstances seem to indicate the probability of its being one of the richest deposits of this region.

The surface of the ground containing these deposits is of a spongy texture, the thin outer coating giving way readily under a horse's feet, and exposing a white powdery substance, formed in great part of sulphate and carbonate of soda. Another characteristic appearance is presented by the lines of small stones crossing the country, resembling the meshes of an enormous net. These appear to me to be due to the shrinkage cracks formed in the soil by drying after heavy rains, whereby it is broken up into a series of polygonal areas elevated in the centre, whence the stones roll into the cracks and fill then up. These lines are considered by the searchers after nitre as probable indications of its presence.

Another circumstance worthy of notice is that the valleys where nitrate is found, have the character of old lake-basins, the greatest thickness of the mineral being found on the slopes; which is probably to be accounted for on the supposition of the formation of the nitrate in lakes, the greatest amount of deposition having taken place round the edge, where the water, being shallowest, would be most rapidly evaporated.

As regards the question of the formation of storage-reservoirs for water, which formed one of the points proposed for investigation by the Honourable Council, I may say that I do not consider that their formation would be specially difficult, as water is known to exist at Cachiyuyal and on the marshes of Cachinal, the first being below and the second above the levels of the nitre-beds. It is very probable that by sinking wells at the points of junction of the lines of watercourse on the bottoms of some of the small intermediate valleys, the subterranean watercourses passing below might be tapped. This question has probably been already settled, because on our return we found steps were being taken for sinking a well in search of water at the second pampa.

If the processes for refining the crude Nitrate should be carried out at the sea-port, as is done in Bolivia, wells would only be required for domestic and culinary purposes by the workpeople employed, and for the accessory operations of this work, the larger consumption necessitated for crystallizing and refining the salt being dispensed with.

It appears to me to be unnecessary to dwell upon the excellent quality of the Nitrate of Cachinal, as this point has been fully demonstrated by the remarkable series of analyses made by Don José Antonio Vadillo, with which the Honourable Council will be perfectly acquainted, as they have been published in their " Annales."

The road from Taltal to Cachiyuyal has a tolerably uniform slope, nowhere rising very steeply. It would be available for the formation of a railway uniting the Nitrate-works to the port; and such a line is, in my opinion, indispensably necessary for the proper development of the works, which will require cheap carriage to be able to compete with the produce of the Antofagasta works, which is by far the most perfect establishment of its class in existence. For although it is true that the Nitrate of Cachinal has the immense advantage of being almost free from common salt, which greatly facilitates the operation of refining, it must be remembered that in manufactures of this kind, where the raw material is only of small value, all the conditions of working, including those apparently the most insignificant, must be so arranged as to reduce the cost of refining to as low a figure as possible.

After our return to Taltal we proceeded to the Bay of Remiendos, which appears to be destined to become the port of exportation of the mineral produce of the region north of Paposo, and of the nitre-works of Aguas Blancas. The harbour is situated at the mouth of the valley of the same name. It is well sheltered from the north, but less completely on the south, the low peninsula on that side being an insufficient barrier against the southerly winds. The anchorage is sufficiently spacious to accommodate eight or ten ships.

The Nitrate-works of Aguas Blancas are situated northeast of Remiendos, and south of Antofagasta, at distances of 22 and 24 leagues respectively. They extend to the north-west of the marshes of the same name, and are bounded on the same side by the extensive beds of salt known as *salares*, below which Nitrate is also found, but in beds of 5 or 6 centimetres thick at most. The greatest thickness of Nitrate is found in the district nearest to the marshes.

I have gone over the concessions held by M. Emeterio Moreno and M. Justo Peña, which are separated from each other by a hill bearing north-east and south-west, and found that the thickness of the bed of calcite or native nitrate exposed in the trial-pits was not, on an average, below 75 centimetres. Three or four of these pits were situate in the last-named concession, and ten or twelve in the former; they extend along a line of about two kilometres, and show the bed of nitrate to be continuous and tolerably compact.

From what I have already stated, we may conclude that the nitre-beds, if not too low in produce (a point that will be soon determined by the analyses of the samples sent to Santiago) give every prospect of a magnificent future. The establishment of colonies of workmen there and at Cachinal will form centres of supply for caravans destined to the more complete exploration of parts of the district now almost unknown, and which probably contain hidden treasures of great importance. These discoveries, it may be hoped, will bring back to our province, formerly so highly favoured by fortune, the life and animation lost through the decline and exhaustion of our former mineral centres.

Copiapo, June 14, 1877.

José R. Martinez.

No. 3.

THE NITRE-BEDS OF THE NORTHERN PART OF THE PROVINCE OF ATACAMA.

Valparaiso, June 19, 1877.

To the Intendent of Atacama, Don Guillermo Matta, *at Copiapo.*

I have the honour to submit with the present letter a

short note on the deposits of nitre in the north of the province under your administration, in fulfilment of the request expressed to me by your Excellency at Copiapo, and would further only desire that you may consider it worthy of publication in the Annals of the Council of Mines. I take this opportunity of addressing to you my most cordial thanks for your kindness in obtaining permission for me to accompany the Government Scientific Commission during the exploration of the district in question, and am, &c.,

Dr. Pedro Sieveking.

Introduction.

The deposits of Nitrate of Soda of the province of Tarapaca, have been known from a comparatively early date; but it was only in 1825 that their working was commenced on a scale of any importance. The native nitrate was at that time sent to Chili to be refined, which circumstance gave rise to the name of Chili saltpetre, originally applied to nitrates of soda in Europe. The exportation of nitrate increased very slowly, and it was only after 1830 that the rapid advance began, which has gone on increasing up to the present time.

The origin of these deposits is an enigma that will probably remain unsolved until a larger number of exact observations than are at present available have been made with a view to determining the point. Nevertheless several theories, of greater or less plausibility, have been propounded by different writers.

Chilian Nitre-beds.

For a long time it was supposed that workable deposits of nitre were only to be found in the Peruvian province of Tarapaca, but with increase in demand and price numerous investigations were set on foot by explorers in the deserts of Bolivia and Chili, for the purpose of finding analogous deposits.

The first of these new deposits were discovered in Bolivia, near Antofagasta, and were succeeded in a few years by others of greater importance in the interior: After this came the opening of the beds at Toco, also in Bolivia. Finally the projectors succeeded in finding in the north of the Chilian province of Atacama undoubted indications of similar beds, with which the present Report is concerned.

The northern coast of Chili rises abruptly from the sea, the wall of mountains being only broken through in a small number of places; and it is by these gorges alone that the plateau can be reached that extends from the foot of the Cordillera to the eastward, and from the latitude of Copiapo to beyond the northern frontier of the Republic.

The climatic conditions, especially that of the almost entire absence of rainfall, and the geological structure of the country, both being analogous to those of Tarapaca, pointed to the probable existence on the plateau of saline deposits similar to those of Peru; and in fact the first of them was discovered facing the port of Taltal. By going up the gorge, which debouches into the bay of Taltal for about eighteen leagues, the waters of Cachiyuyal are reached. At this point the valley widens very considerably, is furrowed on one side by the bed of a stream now dry and covered with a layer of salt, below which, however, at a small depth, tolerably pure water is obtained. About a league further to the eastward the valley merges into the plains, and here at the foot of the hills, which appear to have formed the shores of an ancient lake, deposits of nitrate are found, irregular in character, but locally of considerable thickness. These are encrusted with a variable, but generally considerable quantity of common salt, and are only covered by a thin layer of sand. The rocks forming the surrounding hills are entirely of an igneous character, including granite, augitic and hornblendic porphyries, the granite being in places syenitic. These deposits have been but imperfectly examined, so that we are unable to say whether they rest upon beds of clay or salt, similarly to those of Peru.

In tracing the dry water-course up the gorge, deposits of nitrate are found on either side, which are apparently of a more regular kind than the preceding. The surface is only thinly coated with sand, and the *caliche* (crude nitrate) is of the same kind as the Peruvian. These deposits are not found in the bottom of the valley, but at a certain height above it on the rise of the hills, the former position being occupied by masses of salt.

The whole of the observed phenomena in this part of the desert are very similar to those of Peru, although the examination made is not sufficiently detailed to admit of the establishment of an exact parallel. What may at first sight seem strange, that the nitre should be deposited at the hill-sides, while salt only is found in the bottom of the valley, will cease to appear so when we consider the probable method of formation. We have already said that this formation probably took place in lakes on the sea-shore occasionally invaded by the salt water. If this theory be adopted, it follows that the salts, such as nitrate of ammonia, sulphate of soda, &c., were formed by the decomposition of animal and vegetable matter in the water, while the adjacent rocks similarly gave rise to carbonate of lime, soda, and potash. The mutual reactions of these salts formed nitrate of soda, sulphate of lime, &c., a large proportion remaining in suspension in the salt water. As the lake dried up under the influence of solar evaporation, a gradual concentration of the saline solution was effected, and precipitation took place. As nitrate of soda is only slightly soluble in a solution of chloride of sodium, it might be expected that the separation would go on most rapidly where the concentration of the water was greatest, or near the margin of the lake. We should therefore expect to find the caliche chiefly along the old shore-line, without necessarily supposing it to be absent in the deeper portions, where, however, it might be expected to be found below a covering of salt.

In this theory we have supposed that the surrounding rocks have furnished a portion of the mineral constituents of the deposits; and this appears to be borne out by the facts observed, as we find everywhere large masses of clay gypsum and chalcedony, which can scarcely be other than products of the decomposition of rocks. The extent of the nitre-beds east of the waters of Cachiyuyal does not appear to be less than ten leagues, following the dry watercourse. How far they may reach to the south is not known.

North-east of the waters, the plain is slightly undulated with protrusions of granite rocks similar to those previously noticed. Over the entire extent of this plain, if it can be so called, with the exception of the island-like hills which project at intervals, nitrate is found more or less abundantly, sometimes in beds of as much as three mètres in thickness, and, as a rule, immediately adjacent to the surface without any covering. This enormous deposit, which covers several square leagues, is of a very different character to that described above. The caliche is less mixed with other salts, but in many places it forms a kind of impregnation of the sand or rock without being associated with masses of salt or gypsum, and seems to follow the hollows of the ground. In this case, therefore, the above theory will not meet the facts if the mineral is supposed to be in the position in which it was originally found, but it may still apply if the beds are supposed to be of secondary origin or the remains of deposits formed at a higher level, which have been washed by rains into depressions in the lower ground, where the nitre has been recrystallized by the evaporation of the water.

The northerly extension of these deposits has not been followed, so that we are at present unable to determine the limits of the nitre-beds of Cachinal. In any case it may be supposed that those of Aguas Blancas represent an extension of them. The latter are to the south of the Bolivian frontier, and extend far to the east and south without the limits being exactly known. The explorations made in the neighbourhood of Aguas Blancas cover a tolerably large tract, in which large deposits of nitrate are found, analogous in formation to, and probably extensions of, those of Peru and Bolivia.

From what has been laid down in the preceding pages

it will be seen that the deposits of nitrate already discovered are of considerable extent, and that we may reasonably expect that others will be met with. The total extent of the nitrate-bearing ground is probably not inferior to that of Peru. Detailed and systematic examinations of the deposits discovered have not as yet been possible, on account of the great natural difficulties in the way. These will, however, be probably undertaken before long, when the attention of capitalists is called to them. At the beginning, however, the struggle with these difficulties will no doubt be a hard one, and therefore some time will probably elapse before the manufacture of nitre is permanently established in the province of Atacama, but it probably will be in the end.

Don J. Pedro Sieveking.

Valparaiso, June 22, 1877.

No. 4.

Report of Mr. Domeyko on samples of Caliche from Atacama.

Analysis of two pieces of nitrate forwarded in January last to the Minister of the Interior by Don Saturnino Corvalan, and obtained from a nitre-district discovered by him in the Chilian portion of the desert of Atacama, which have been analyzed by order of the Minister of Finance.

44

Santiago, April 11, 1877.

M. LE MINISTRE,

Conformably to the commission received from your Excellency by letter on the 20th of March last, I have analyzed the two large fragments of native saltpetre received at the same time, with the following results :—

1. The largest mass is of an imperfectly homogeneous character, being streaked with white veins of nearly pure common salt, and others more charged with nitre. The principal portion is of a crystalline texture mixed with earthy matter, and has the following composition.

Nitrate of soda............................	24·30
Chloride of sodium (common salt) ..	53·65
Sulphate of soda	4·95
Insoluble earthy matters................	9·80
Lime and magnesia.......................	traces
Water and moisture......................	7·30
	100·00

2. The second smaller piece is more homogeneous, and of a pure-grained texture. It contains :

Nitrate of soda	27·98	30·00
Chloride of sodium...................	23·00	
Sulphate of soda	8·46	
Sulphate of lime (gypsum)	3·41	
Magnesia..............................	traces	
Insoluble earthy matter	14·70	
Water	22·45	
	100·00	

The two samples are very similar both in composition and external appearance to the *caliches* of Peru and of the nitre-beds of Carmen, near Mejillones.

IGNACIO DOMEYKO.

No. 5.

Report of Mr. Domeyko on the Guano of Atacama.

Santiago, June 25, 1877.

M. LE MINISTRE,

I have carefully examined the sample of the Guano recently discovered in the desert of Atacama, and with the analysis of which I have been entrusted by your Excellency's dispatch of the 14th of May last. The result of this examination shows that this guano contains a large quantity of nitrogen, and resembles in composition that of the Chinchas Islands, being very different from the phosphatic guano of Mejillones. The odour is very fetid, the colour is dark brown, and the mass, which is light and porous, consists in great part of fecal matters and ammoniacal salts. When calcined, it leaves behind 15 per cent. of incombustible, non-volatile residue. By assay it gives 12 per cent of nitrogen, which is equivalent to 14·4 per cent. of ammonia, or nearly as much as in the good guano of the Chinchas; but the phosphoric acid does not exceed 4·3 per cent. Of this a small portion only is present as phosphate of lime, the proportion being scarcely 1·6 per cent., the remainder being in the state of soluble phosphate, which is considered by agriculturists as a more active manure than phosphate of lime.

The above is all the information that I am able to furnish to your Excellency concerning the matter.

IGNACIO DOMEYKO.

To the Minister of Finance.

Observations of the Official Journal on the scientific valuation of the Guano of Peru.

SHORT COMPARISON.

It may not be without interest on the occasion of the publication of the first result of the analyses of the newly discovered Guano of Atacama, to present to our readers those made in France by M. Barral of the Peruvian guanos of Macabi and Guañape, which, according to that chemist, as well as other eminent authorities, were as good and rich as that of the Chinchas.

The following are three analyses taken, with their dates, from the 'Journal de l'Agriculture.'

1. Cargo landed in a French port January 21, 1874.

Water.............................. 30·19
Organic matter and ammoniacal salts 36·45 = Nitrogen 10·01
Phosphoric acid .... 13·23
Lime, potash and soluble substances 18·62
Insoluble substances .... 1·57

100·00

2. Cargo of 1st April, 1874.

Water 28.40
Organic matter and ammoniacal salts 36.44 = Nitrogen 10.10
Phosphoric acid........................ 15.17
Lime, potash and soluble substances 17.97
Insoluble substances.................. 2.02
                                      ———
                                     100.00

3. Cargo of June 20th, 1874.

Water 28.60
Organic matter and ammoniacal salts 40.00 = Nitrogen 11.38
Phosphoric acid........................ 13.24
Lime, potash and soluble substances . 16.46
Insoluble substances.................. 1.70
                                      ———
                                     100.00

As a general average, the Guano of Guañape contains, according to M. Barral, nitrogen 10—12 per cent.; phosphoric acid 12—15 per cent., and potash 2—3 per cent. On comparing the above figures with those of the analysis made by the Rector of the University, we find that the Chilian guano is richer in nitrogen than the Peruvian guano of Guañape, which was considered by the most eminent agricultural chemists of Europe, as well as by the farmers who employed it, to be equal to the famous guano of the Chinchas islands.

The statement made by the Rector of the University that the guano analysed by him " contained nearly as much nitrogen as the good guano of the Chinchas," must be considered as a guarded statement necessarily made in a purely scientific work.

No. 6.

COMMISSION CHARGED WITH THE EXPLORATION OF THE NORTHERN COAST-LINE OF ATACAMA.

Report of the Commander of the Abtao.

Antofagasta, November 25, 1876.

To the Naval Commander-in-Chief,

Conformably to the communication which I had the honour of addressing to your Excellency on the 26th of last month, after having embarked the engineers, Don Eugenio Plazolles and Don Macorio Sierralta, and in accordance with your instructions to proceed to an examination of the Atacama coast-line on the 24th parallel of south latitude, in the small harbour of Cobre; I left the port of Mojillones on the day mentioned above, and anchored at Antofagasta in the night of the 27th October, 1876, where the engineers above mentioned completed the necessary stores and equipment required by the expedition, and a person was embarked having a good knowledge of the ground and the valleys leading to the interior of the desert. This was Don Segundino Corvalan, an old Chilian miner, now domiciled at Antofagasta, who was good enough to render, gratuitously, services of the highest value towards the successful carrying out of the mission with which I was entrusted. I therefore considered it my duty to bring these services specially under the notice of your Excellency. Having completed the preparation, I left Antofagasta on the morning of October 31st, under easy steam, and proceeded southwards within a distance of a mile from the shore. At 9·40 A.M. I arrived opposite the stone pyramid marking the boundary of Chili and Bolivia on the 24th parallel. The pyramid, which was plainly visible at the short distance (about three-quarters of a mile), is placed upon a rock off the shore about 20 metres in height. As it was at this point that the detailed examination was to be commenced, I stood along the shore, as close in as it was possible to do without endangering the safety of the ship under my command. About a mile and half south of the boundary-line the mountains are divided by a tortuous valley with steep sides, making an undulation in the chain that may be useful as a guide in finding the frontier point from a distance.

The coast extends southward in a nearly straight, or but slightly inflected line, none of the bays being sufficient to shelter a vessel of any size, to within seven miles of the 24th parallel, where the small harbour, called by the miners and mineral proprietors of the desert, Agua Dulce, is situated, and where I anchored on the same day at 11 A.M. This place, situated in lat. 24°·07' S. and long. 72° 32' 20" W. from Greenwich, although sheltered from winds between azimuth of south and south-west at most, is not protected against the swell setting in from the latter point, which circumstance renders the anchorage difficult, the shore being without beaches, and having a rocky bottom. The harbour is surrounded by steep mountains, averaging about 2000 feet in altitude to the south, which culminate in a prominent peak, known as Agua Dulce, about 3000 feet high, which serves as a guide-mark to vessels entering. The water is deep, the bottom being of stones, sand and shells; the best mooring-ground (marked by an anchor in the plan) is in 20 fathoms; bottom sand and shells; bearing, on the starboard hand, a white rock on the south shore. There is good anchorage for smaller vessels in 13 fathoms, on a sandy bottom to the eastward of the above point.

Two lines of valley meeting at the shore penetrate the mountains at this place; neither of these, however, when examined by the civil engineers on board, proved to be fit for a road to the interior. There is also a want of level ground or beaches suitable for buildings, the shore being rough and rocky all round. We remained at this place for one week, and after having made a detailed survey of the port on the scale of $\frac{1}{10000}$, and of the coast-line to the 24th parallel, sailed on the 1st of October, and anchored the same day at Agua Salada, four miles further south, in lat. 24° 11′, long. 70° 32′ W. and four miles north of Cobre, a harbour which is somewhat better than that of Agua Dulce, being larger, and having a better bottom, but similar in character as regards shelter from prevailing winds. There is room here for 15 or 20 ships; and the place may be easily found by a yellow mark covering the mountains on the north for about two thirds of their height. The best anchorage is in 12 fathoms, on sand and stones, at the point indicated in the chart. The shore is rough and steep, and offers the same impediments to buildings as that of Agua Dulce.

At the bottom of the bay, on the south-east side, a gorge comes down to the shore, and a short distance up joins another known as Lobo Muerto, which communicates with the open roadstead of the same, south of Agua Salada.

The examination of this valley made by the civil engineers led to the following conclusion. It would be possible by constructing a waggon-road up the side of the first gorge, and then through the second, to open up communication between the port and the central valleys of the interior; this, however, would be a difficult and expensive work, the details of which will, however, be gone into by the engineers entrusted with the topographical examination through the proper official channel.

After completing the necessary hydrographical surveys, and uniting them with those of the north, I started again on the first of this month, and arrived the same day in the harbour of Del Cobre, where my commission terminated. Having, however, been informed at Antofagasta by credible persons having special knowledge of this part of the desert, that there was an apparently good harbour, about eight miles south of Cobre, known as Remiendos, situated at the mouth of a valley, which appeared, from its slope and other characters, to promise well for a line of road to the interior, and their information being confirmed at Cobre, I resolved to examine the new harbour. On the 10th therefore I left Cobre, and two hours and a half later anchored at Remiendos.

This harbour, in lat. 24° 26′ S. and long. 70° 35′ W., is sheltered from south and south-west winds, and although of small area is capable of sheltering 12 or 15 vessels. The water is not very deep; the bottom is of fine sand, with the exception of a few rocky patches. The harbour is formed by a peninsula on the south side, nearly circular in plan, about 900 metres in diameter, rising 15 metres above the sea-level. From its north end a line of reefs extends for about 210 metres; all of these are above water, with the exception of that at the northern extremity, which is in 2 feet water at the lowest tides. A heavy sea was breaking upon it constantly during the week of our stay. The natural breakwater is about 10 metres in extent, and is free from the sea-swell which is abundant in the eastern part of the harbour, where there is a large rock 200 metres from the shore.

This place is well surrounded by level ground suitable for buildings, and sheltered places fit for quays and for mooring boats and lighters, the sandy beach at the bottom of the bay forming an excellent landing-place. Towards the north, about a mile from the anchorage, is the mouth of the great valley of Remiendos, which leads in a nearly easterly direction to the interior for about 40 kilometres, where by a short pass it communicates with the central valleys of the desert. This ground has been carefully examined by the engineers, who estimate the average slope to be from 4 to 5 per cent., and think that it is in every way suited for the projected line of road.

The harbour and valley of Remiendos, being situated between two celebrated mooring-localities, like those of Cobre and Paposo, command a rich and extensive district, containing, as I am informed by persons having a good knowledge of the country, an abundance of copper- and silver-ores, which are at present inaccessible, owing to the immense difficulties encountered by the explorer.

Having terminated the hydrographic surveys at this point, on the 17th I left for Antofagasta under sail, where I arrived on the 18th, and discharged Messrs. Plazolles and Sierralta, who proceeded to their destination.

The general survey of the coast, from the 24th parallel of S. latitude, as far as Remiendos, inclusive, and detailed plans of the different harbours, will be sent to your Excellency as soon as they are completed; meanwhile I append to this report a sketch survey of the harbour of Remiendos.

The hydrographical surveys in question have been carried out under my direction by first lieutenant Don Louis Lynch, with the assistance of the midshipmen Don Adolf Castro and Don L. Fierro, and I have much pleasure in bringing to your Excellency's notice the zeal and application of these officers.

I am, &c.

T. RONDIZZONI.

No. 7.

Report of Messrs. Plazolles and Sierralta.

Valparaiso, 28th November, 1876.

The Minister of the Interior.

The war steamer 'Abtao' having terminated the exploration of the coast of the desert of Atacama, between the 24th parallel of south latitude, which marks the boundary between Chili and Bolivia, and the Harbour of Remiendos 27 miles further south, I have the honour to report to your Excellency upon my part of the expedition, especially that of the exploration of the land which has been entrusted to me, conjointly with Don Sierralta.

On our arrival at Mojillones from Bolivia, on the 26th of October, we placed ourselves at once under the orders of the commander of the 'Abtao.'

We found the Chilian inhabitants of Antofagasta to be well disposed towards and desirous of aiding the expedition; and one of them, Don Segundino Corvalan, an old friend of the engineer Sierralta, who has been for many years interested in mining affairs and accustomed to travel in the desert, readily consented to accompany us. The valuable cooperation of this gentleman was attended not only with a considerable saving of time, but also spared much of the suffering which all must experience who have not been previously accustomed to living in the desert region.

Having completed our preparations, the 'Abtao' left Antofagasta on the morning of the 31st October and commenced a minute examination of the coast-line from the boundary-monument between Chili and Bolivia, about $3\frac{1}{2}$ kilometres north of the 24th parallel.

Harbour of Agua Dulce.

The first anchorage entered was that of Agua Dulce or, as it is sometimes called, Port Montt, situated 8 miles south of the 24th parallel. Here the hydrographical surveys were at once commenced, being entrusted by the Commander to the Lieutenant Don Luis A. Lynch, with whom we had great pleasure in cooperating during the spare intervals in the land surveys. I do not propose to enter into points of hydrographic and nautical detail, which will be done completely and authoritatively by the Commander Don Francisco Rondizzoni, but will confine myself, in accordance with my instructions, to a general view of the question as to how far this harbour is suited for the establishment of a fixed settlement and for the starting-point of a road to the interior of the desert.

Such parts of the harbour as are protected from the south-west winds, and therefore suitable as an anchorage, are surrounded by a chain of mountains which rise steeply from the southern point to a height of 1200 feet, directly from the sea, without leaving any level ground suitable for a town site; and even the small number of houses that it might be possible to erect would be constantly liable to injury from the slipping of the loose material forming talus on the hill-sides.

By following the shore to the northward for about three miles inland from the harbour a plain sufficiently large and suitable for a town site was discovered. This extends northwards by the valley of Agua Dulce, the dry bed of an old watercourse. The road to the place could only be made fit for vehicles at a very considerable cost. In the sheltered parts of the harbour boats might be moored, and a wharf could be built at small cost; but it would be 3 miles away from the town. In looking for a more favourable combination of positions, we found, 2 miles to the north of the main harbour, a beach sheltered by a small island from the south-westerly sea. If this were united to the mainland by a pier, which could be easily and cheaply constructed, a small shelter harbour for boats and lighters would be formed; but even in this case the landing-place would be about a mile from the town site.

These conditions, though not very favourable, might be accepted if the locality gave prospect of a ready means of communication with the interior of the desert. We found, however, that the gorge of Agua Dulce was far from fulfilling this requirement, as it would be impossible to carry a practicable waggon-road through it except at enormous cost; and even then the gradients would be excessively steep in places (up to 1 in 10). Furthermore, the northerly trend of the gorge renders it extremely possible that it may pass into Bolivian territory; but this point we did not attempt to determine, the proposed solution of the problem by this route being otherwise inadmissible.

Harbour of Agua Salada.

After completing our examination of the first harbour, we proceeded in the 'Abtao' to the next, situated 4 miles further south, and known as Agua Salada. This is in latitude 24° 12′ S., and presents a tolerably good anchorage, sheltered from the south-west. It is smaller than Agua Dulce, and is similarly surrounded by high mountains, broken through on the south side of the harbour by a depression through which the gorge of Lobo Muerto might be reached by a road from the sea-shore along the slope of the mountains. But there is no level ground suitable for a settlement, and the road would be very costly and, in many places, excessively steep. The gorge of Lobo Muerto comes down to the sea on an exposed part of the coast, where there is not the slightest shelter for vessels.

HARBOUR OF COBRE.

From the harbour of Agua Salada the 'Abtao' proceeded to that of Cobre, situated about 15 or 16 miles south of the 24th parallel.

This place, which has been for some time exclusively in the occupation of the firm of Barazarte, formerly Moreno, cannot be put into communication with the interior of the desert by a waggon-road at a reasonable cost, and is without sufficient ground to accommodate a population of any extent. Up to this point, therefore, we were unable to congratulate ourselves upon having found any thing like a satisfactory solution of the problem; but having been informed of the existence of a good harbour a few miles further south, and a valley suitable for the formation of a carriage-road, the 'Abtao' proceeded thither, arriving at the anchorage on the 11th of November.

HARBOUR OF REMIENDOS.

The Harbour of Remiendos, of which a plan is attached to the present Report, may, notwithstanding its small size, become a place of considerable maritime importance. It has an excellent sandy bottom, and a peninsula of about 100 acres in area, with a mean diameter of 700 mètres, protects the anchorage on the south-west. Along the line A B (marked in red ink on the plan) a quay might be built along the shore in a perfectly sheltered position, alongside of which the cellars and warehouses of the Customs would be placed. The direction of the jetty is from S.E. to N.E.,

E

and by making it 200 metres long, large vessels could come alongside, there being sufficiently deep water at the north end. The surface of the peninsula has a mean elevation of 8 to 10 metres above sea-level; the isthmus connecting it with the mainland extends from N. to S.E., and is 400 metres long by 100 metres broad. The surface, which is covered with fine sand, is only elevated 2 or 3 feet above high tides; but it might easily be protected by quay walls, or the hollow might be filled up by an embankment of a metre in height, the materials being ready to hand if a few undulations of the neighbouring ground were levelled. The ground being well situated near the pier, and in the centre of the proposed town, would be among the first portions to be taken up; and therefore the Government might make the execution, of these embankments, at the cost of the allottee, part of the condition of the concession of these lots.

As there is a fine sandy beach on the north of the isthmus suitable for beaching boats and small vessels, it would not be absolutely necessary to construct a quay for landing passengers and baggage.

West of the isthmus the flat ground follows the coast, and extends to the base of the mountains, forming good building-sites, with an average length of 6000 feet and a breadth of 2100 feet. There is therefore in all nearly 300 acres of ground, including that upon the isthmus and peninsula, suitable for a town-site within a reasonable distance of the harbour. The lie of the ground is such that in laying out the town the principal streets should follow the magnetic meridian approximately.

It is therefore evident on a consideration of the preceding details that Remiendos presents all the elements necessary for the establishing of a town susceptible of considerable development. There will be no want of space for the industrial establishments that will be required should the working of the nitrate-deposits in the interior be actively undertaken.

The coast mountains falling back a little to the eastward leave the proposed town-site fully exposed to the beneficial southerly winds, so that the air is fresh and comparatively cool.

A short distance to the south of Remiendos, about 2 or 3 miles following the coast, are the waters of Botigas. These are brackish, although it is said that beasts of burden drink them without inconvenience. I do not consider that they would be of any use for the town supply, which would be obliged to have recourse to the distillation of sea-water to obtain a supply for drinking.

About $3\frac{1}{2}$ kilometres from the jetty, following the coast to the north, is the entry of the valley of Remiendos, the dry bed of an old river, along which for 40 kilometres the waggon-road could be carried.

The mouth of the valley is barred by a spur of rock about 20 metres high, from which point there are no obstacles for 18 kilometres. The next 5 or 6 kilometres are steep and rocky, and would require some considerable amount of earthworks for the establishment of a good road of a minimum breadth of 6 metres. From this defile the valley opens in every direction; but it would be preferable to follow the old river-bed up to the pass, the altitude of the latter point being 1753 metres or 5750 feet above the sea-level.

The above line may be divided into four sections, which will now be considered more in detail.

First Section of the Waggon-road.

The first section is 3500 metres long from the jetty to the upper part of the spur of rock at the mouth of the valley, which is 250 feet, or 76 metres, above the sea-level, showing an average rise of 1 in 50, if the jetty is assumed to be 6 metres above the sea-level. The road might form part of the town street for $1\frac{1}{2}$ kilometre, when it would be carried up the slope of the hills so as to rise to the top of the spur. This would require a length of about 500 metres of cutting in rock.

Second Section.

The second section, of 18 kilometres, presents no special difficulties. The greater part of it being, however, in the sand and gravel of the old river-bed with occasional large stones, I should recommend that the latter instead of being thrown on one side, should be laid in a kind of pavement, the sandy soil being very soft and never being wetted by rain, and therefore likely to wear under the pressure of wheels to a considerable depth. In order to save expense, this pavement should be laid in three lines of about a metre in breadth, more or less parallel to each other in the centre, and at either side of the roadway. In any case it would be advantageous in traversing the loose and incoherent ground in the desert to make use of wheels with tires 20 to 25 centimetres in breadth.

The second section, starting from a height of 76 metres,

reaches 915 metres, the difference between the ends being 839 metres, giving an average slope of rather more than $4\frac{1}{2}$ per cent., which appears to be tolerably uniform over the entire length.

Third Section.

The third section of 6 kilometres is, as we have said, the most difficult and costly part of the road. The ground being cut up by gorges and barred by escarpments of rock, I consider that the cost of breaking a road 6 metres broad through it would be from 25,000 to 30,000 dollars. The height of the lower end is 915 metres, that of the upper end 1281 metres; the difference of 366 metres, when compared with the length, 6 kilometres, gives an average slope slightly in excess of 6 per cent.

Fourth Section.

The fourth section of 13 or 14 kilometres passes through an open country, and is that most favourably situated. The road-bed is clean but soft, an inconvenience that must be met by the use of broad wheels. The height of the lower end is 1281 metres, that of the terminal point 1753, the difference of 472 metres being equal to an average slope of about $3\frac{1}{2}$ per cent.

Approximate Estimate of Cost of Road.

I have estimated in the following table for a road of 6 or 7 metres broad, sufficient for two carts to pass easily, and properly levelled and paved with stone wherever it can be got on the spot. Under these conditions, the separate sections would cost as shown below :—

ROAD IN THE VALLEY OF REMIENDOS.

Section.	Length.	Lower end.	Upper end.	Difference of level.	Average inclination.	Cost per kilometre.	Cost of section.
	Metres.	Metres.	Metres.	Metres.		$	$
1	3,500	6	76	70	2 per cent.	3000	10,500
2	18,000	76	915	839	4½ „	2000	36,000
3	6,000	915	1281	366	6 „	5000	30,000
4	13,500	1281	1753	472	3½ „	1000	13,500
	41,000						90,000

This estimate will not be considered excessive when it is remembered that the workmen must be supplied with drinking water while at work, and that they will require somewhat higher wages than those current in inhabited regions. The figures given include all general expenses, such as the cost of passage of workmen, temporary lodging, &c.

If it were desired to limit the works to what is indispensable for making the route passable for carts, it would be sufficient to carry out the earthworks of the first and third section, and pick out the loose stones covering the line of the second section, throwing them to the right and left, but without making any earthworks either in this or the fourth section. Under these conditions I believe that the outlay need not exceed $40,000.

At the head of the pass the ground descends to the eastward in a gentle slope to a vast plain resembling the bed of an old lake. Further on is a line of gently undulating height, extending north and south, behind which are the nitrate beds of Punta Negra and Profeta. Those of Aguas Blancas, in about 24° south latitude, may be about 15 leagues to the north; but all these parts of the interior are easily accessible by carts from the pass at the head of the valley of Remiendos.

The main valley is joined at intervals by secondary gorges practicable for carts, which will be of value in the working of the minerals existing in this part of the coast-range. Having obtained the information detailed in the preceding paragraphs, we have considered our mission to be accomplished. On my part, I desire that your Excellency may find in my Report the information required for realizing a work which, as far as I can judge, will be of great advantage to the country.

E. PLAZOLLES,

Civil Engineer.

P.S.—Mr. Sierralta is in possession of the field-notes of observations made on the ground with a pocket compass and aneroid barometer; the distances were estimated by a mule's paces. From these a sketch map will be made for your Excellency's information.— E. P.

On board the 'Santa Rosa.'

29 November, 1876.

Having been honoured with the mission of inquiry into

the means of connecting the desert with a seaport as near as possible to the Bolivian frontier by a carriage-road, I have great pleasure in reporting to your Excellency the results of a minute investigation of the subject.

On arrival at Mejillones, and after embarcation on board the 'Abtao,' the commander, Mr. Rondizzoni, returned to Antofagasta, where Don Segundino Corvalan generously volunteered to accompany us, together with MM. Raphael, Sutil, Borjas Besoain, and Francisco Basenñan.

We left Antofagasta and proceeded south, stopping at the harbour of Agua Dulce or Port Montt, which is about 4 miles south of the 24th parallel. This place has a good anchorage, sheltered from southerly and, in part, from south-westerly winds, and can accommodate fifteen vessels safely without difficulty. To the north of the harbour is a gorge which I did not explore, as it enters Bolivian territory within a short distance. The water is tolerably smooth, the depth all round the bay, at 80 metres from the shore, is from 4 to 5 fathoms.

Upon the whole shore of the bay there is nowhere sufficient ground for the formation of even a moderate-sized town, the only ground at all likely for the purpose is at the mouth of a gorge noticed below; but this is at too great a distance from the harbour.

The gorge in question is situated about the middle of the bay, and was explored by me, together with Mr. Orella, lieutenant of the ship. It is traversed by escarpments of rock, some of which are 7 metres in height. By dint of hard work we managed to ride up it for a distance of 10 kilometres, when I was obliged to give up further trial in this direction.

The observed distances and the altitudes determined by the barometer convinced me of the impossibility of carrying a waggon-road through this gorge, as the mean slope would not be less than 1 in 10, and even this would require enormous cuttings and embankments, with a correspondingly heavy cost.

Furthermore, Mr. Corvalan pointed out that the difficulties were still greater at a more distant point; and this, together with my own observations, induced me to abandon the investigation and return on board the 'Abtao' to report the result to the Commander, who agreed with me in considering the exploration of this point at an end. The results obtained not being satisfactory, we continued our voyage southward and arrived at the harbour of Palo Varado, which is 7 miles south of the preceding, and 1½ mile north of the lower end of the gorge of **Lobo Muerto**, where the conditions are somewhat similar to those already described, so far as regards building-sites, anchorage, and shelter from prevailing winds; but the difficulties of road-making were sensibly less.

Two gorges come down to the shore at this place. One of these soon comes to an end on a high mountain. The second and southern one leads to a defile, by crossing which the gorge of Lobo Muerto is reached. From this defile the road is practicable to another, situated 26 kilometres from the coast, and 4750 feet above the sea-level. This can be passed without difficulty, and leads down to the gorge called Mateo, which is on the line of the nitre-beds of Aguas Blancas, and branches into that opening a little south of the Port of Antofagasta, along which the line of the Antofagasta Nitrate Company's railway is carried.

The gorge of Lobo Muerto has several rocky barriers, three being remarkable as exceeding $7\frac{1}{2}$ metres in height. But these are not the only difficulties in the way of road-making, there being narrow passages that would require blasting on a large scale, and long distances obstructed by loose stones and projecting rocks, which render the way difficult even for mules. A considerable amount of filling would also be required to reduce the slopes, which are very steep.

The above statement refers to the portion of the road included between the junction of the gorge of Lobo Muerto with that bearing from the harbour. The which of the two parts of these gorges is to be chosen is in no way doubtful, as I shall proceed to show.

1. The observations made upon the gorge of Lobo Muerto from the point of junction spoken of above to the coast were obliged to be made on foot, because no mule could be induced to go down it. The distance is 6 kilometres, upon which 12 escarpments were found, some being 15 metres high. The road is impracticable in this direction; for in some places the incline would not be less than 15 per cent.

2. It now remained to examine the gorge which, starting from the harbour, passes by the first defile. In order to carry out the road it would be necessary to commence it about 2 kilometres north of the opening of the gorge, and to carry it on the slopes of the mountain for 2 or 3 miles, to pass the first defile, and afterwards to unite it with the gorge of Lobo Muerto by a similar piece of work, which would only need to be 1 kilometre long. This line is preferable, and would not present any insurmountable diffi-

culties in the event of a road being required at this part of the coast. Notwithstanding the difficulties indicated, a practicable carriage-road might be constructed with an average incline of 6 or 7 per cent.; but it would be a costly work, the execution of which could only be justified in the event of no better line being found.

In this investigation I was accompanied by MM. Orella and Corvalan.

The work being finished at this point, I informed the Commander of the result, who determined upon continuing his voyage towards the Port of Cobre.

The harbour of Cobre is a bad one, being exposed to the prevailing southerly and south-westerly winds, as well as to the north. It is small and dangerous both to vessels entering and leaving; the sea runs so high that sometimes for days together they can neither load nor discharge cargo. It may also be said that a carriage-road to the interior is an impossibility : there being no natural depressions, it would be necessary to cross a series of high hills at great cost in order to reach the gorge of Lobo Muerto. I shall, therefore, say nothing more about this place.

At the port is an establishment belonging to the executors of M. J. A. Moreno, in which some important copper-mines are included.

After making the above observations, and seeing that it was useless to remain any longer at this place, we proceeded to Remiendos, which, according to our information, was the only place likely to fulfil all the desired conditions. This harbour, which has furnished us with a sufficiently satisfactory solution of the problem, is situated 10 miles south of that of Cobre, in latitude $24° 26' 59''$ S, on the same parallel as Aguas Blancas and the other nitre-beds in the interior.

The area of level ground at this place is sufficiently large to accommodate a very considerable population.

In the south part of the harbour is a fine clean sand beach, 60 metres broad, with smooth water. At this point the sea forms a bay capable of containing 200 small vessels, and there are convenient sites for lighthouses, quays, and public buildings. At a small cost a good wharf of 3000 metres long might be built along the shore; and by making it 30 metres broad there would be a depth of 5 fathoms alongside. At the south end of the harbour the rocks projecting into the sea might be united by masonry work in order to increase the natural shelter.

The south-westerly winds that penetrate into the harbour are usually very much reduced in force. Between southwest and north-west the harbour is exposed to the winds, but in all other directions it is sheltered.

The bottom slopes gently to a depth of 25 fathoms without steep ridges. The 'Abtao' was anchored in 9 fathoms; and although the wind blew hard from the south for two days, she rode it out safely at single anchor with 40 fathoms of chain. The bottom is of sand, shells, and very little stone, sand prevailing over the principal part of the harbour.

Ships may leave with the wind from three out of four quarters of the compass; the anchorage is large enough for 20 vessels. The southerly wind, which is most prevalent, does not raise any very great amount of sea, and it offers no hindrance to vessels either entering or leaving.

There are no reefs in the harbour placed so as to render

the navigation dangerous, there being only a single rock below water at the extremity of the peninsula; but this is perfectly well marked. On the north shore, as far as the valley, are several small islands, which would afford sites for foundries, smelting-works, or other establishments of a similar kind. About 2 miles from the bottom of the harbour is the valley of Remiendos, which is suitable for a carriage-road. This would present some difficulties, but they might be easily overcome. At the commencement of the valley is a scarp of 20 metres in height. To avoid this the road would be carried on a gentle rise along the slope of the mountain for two miles; 500 metres of this is in very hard rock, but one that breaks easily when blasted. After passing this first obstacle, no others are met with for 14 kilometres, when the principal difficulties commence. In a length of 3 kilometres numerous scarps, large rocks, and narrow gullies must be passed; and in this section the road would be at an incline of 5 or 6 per cent., the maximum slope permissible in a carriage-road, and the greater part of it would require blasting. Beyond this point the work presents no difficulty of any kind up to the pass from which the interior plain is seen. From this pass to the nitre-beds of Aguas Blancas the distance is about 16 leagues, and the road might be carried there by three different lines without encountering any kind of obstacle. The height of the summit pass is 5750 feet, and the distance from the shore 40 kilometres, corresponding to an average slope of 4 or 5 per cent. I consider the cost of a carriage-road of a sufficient width on this route would be about $100,000. The promontory, situated on the south side of the harbour, is from 800 to 900 metres broad, and 15 metres high above the sea-level.

There can be no doubt as to the positive advantages presented by this route for the realization of the project of a line of communication between the coast and the interior, or as to the facilities which it offers to the intrepid explorers of the desert, who have not spared any sacrifice in order to bring to the notice of the commercial world its immense and rich deposits of nitre, and its mines of silver and copper. It will open a vast field to the labour of thousands of industrious hands who are now driven to seek a hard livelihood in a foreign land, and will add an important contribution to the productions of our country.

Time was wanting to allow of my visiting either the mines in the neighbourhood of the line of country examined, or the nitre-beds, which will find an easy outlet for their produce in Remiendos; but I know from information derived from persons well acquainted with the localities that there is a sufficient field for a large development in the direction of copper- and silver-mining.

MACARIO SIERRALTA,
Civil Engineer.

No. 8.
Decree opening the Port of Remiendos (now called Blanco Encalada).

Santiago, April 11, 1877.

In sight of the preceding note, the harbour called Remiendos is opened to commerce as a minor port, and depending upon the Custom House of Chañaral de Animas.

Let the present decree be registered, communicated to those who have right to know it, and published.

PINTO,

RAFAEL SOTOMAYOR.

Decree ordering the establishment of centres of population at Taltal and Blanco-Encalada, and determining the positions of one and other places.

Santiago, June 26, 1877.

In sight of the plans annexed of the port explored in November, 1876, by the steamer 'Abtao,' at the promontory Blanca, situated north of the point called Remiendos, and also of the port of Taltal, and in conformity with the law of November 21, 1876, which authorizes the formation of centres of population at the ports opened by the Customs of the Republic.

I have resolved and decreed:—

Art. 1. There shall be established in the port recognized in the said promontory *Blanca* a centre of population, which shall be called Blanco Encalada, and another centre of population at the port of Taltal.

Art. 2. The town of Blanco Encalada shall be composed of twenty-three blocks, surrounding the ground reserved for the public plaza, of which twenty shall be squares of 20 metres wide, two shall be 100 metres long and 40 broad, and one of 100 metres by 50 metres, separated one from the other by streets 20 metres broad.

Art. 3. The town of Taltal shall be composed of eleven blocks, nine of which shall be 100 metres square, and two 100 by 50, arranged on both sides of the public plaza, and separated by streets 20 metres wide.

Art. 4. At Blanco Encalada the blocks numbered 22 and 23 shall be reserved for public buildings, and at Taltal those numbered 9 and 10 shall be reserved for the same purpose.

Art 5. In either town each of the remaining blocks shall be divided into eight lots, of which six shall have 33 metres frontage and 40 metres depth, and the remaining two 20 metres frontage and 50 metres depth.

Art. 6. These lots shall be conceded to those persons who may ask for them, and who shall bind themselves by a duly authenticated instrument to enclose them within six months of the date of concession, and to build upon them within two years of the same date.

Art 7. In the event of the above conditions not being fulfilled the concessionaires will lose all right over the lots, which will be realloted to such new colonists as may ask for them, and any improvements will remain to the profit of the State.

Art. 8. The Intendency of Atacama will take the necessary measures for the execution of the present decree, and will observe the following rules in making the allotments.

1. Two lots may not pass into the possession of one individual whether by concession or transfer.

2. Preference in allotment is to be given in one or other of the ports in the order in which the demands have been made, and to the coast colonists who have been put out of work by the earthquake of May last.

3. The title to the concession is a provisional one, which will be exchanged for a definite one, on the completion of the conditions announced in Article 6 of the present decree.

Art. 9. Demands of concession of lands for the establishment of smelting-works or saltpetre-refineries must be presented to the Intendant of the Province, who will make the concession conditionally until the law has been otherwise stated, observing the following conditions :—

1. Every concession shall be provisional, and in force until the law has otherwise decided.

2. Only as much land will be granted as will, in the opinion of experts, be strictly necessary for the purpose of the establishment which it is proposed to found.

3. Every foundry or calcining-works must be placed at a distance from centres of population, in such a manner as not to injure the salubrity of the air.

Let this decree be registered, communicated to whom it may concern, and inserted in the Bulletin of Laws.

PINTO,

JOSÉ VICTORINO LACTARRIA.

F

No. 9.
COAST OF CHILI.
HARBOUR OF BLANCO ENCALADA.

Art. 137, 1877. The commander of the iron-clad 'Blanco Encalada,' Captain Don Juan Lopez, announces that he has determined, by good astronomical observations made on shore, the position of the harbour of Blanco Encalada, and has obtained the following results for the landing-place where the sandy beach meets the first rocks to the west.

Latitude south, 24° 22' 20",
Longitude west, 70° 36' 51".

The remaining details furnished by the documents of the Report confirm those contained in the Hydrographical Notices, Nos. 2 art. 9 of January 16th, 1877.

English chart 4277, Chilian chart 10.

HYDROGRAPHICAL NOTICE.
COAST OF CHILI.
HARBOUR OF BLANCO ENCALADA.

Art. 151, 1877. In consequence of a decree of the Government of the Republic, authorizing the establishment of a centre of population in the harbour of Remiendos on the coast of Atacama, it has been thought necessary to change the name of this harbour, to avoid confusion, and qualifying it as an inhabited place, which it soon will be, by calling it Blanco Encalada. As, furthermore, this point of embarkation must soon acquire a considerable commercial importance, on account of the Guano and Nitre found in the neighbourhood, the Hydrographic Office has thought it necessary to adopt the name of Blanco-Encalada for the harbour, where a centre of population of the same name is to be established. Consequently, and as a means of contributing to the technical unity which it is essential to preserve in universal technology, the Chilian hydrographic Department will not recognize any other name for this harbour.

This notice concerns the Hydrographical notices No. 2, Art. 9, and No. 27. Art. 137, of 1877.

English chart 1277, Chilian chart 10.

Francisco Vidal Gomez.
Director

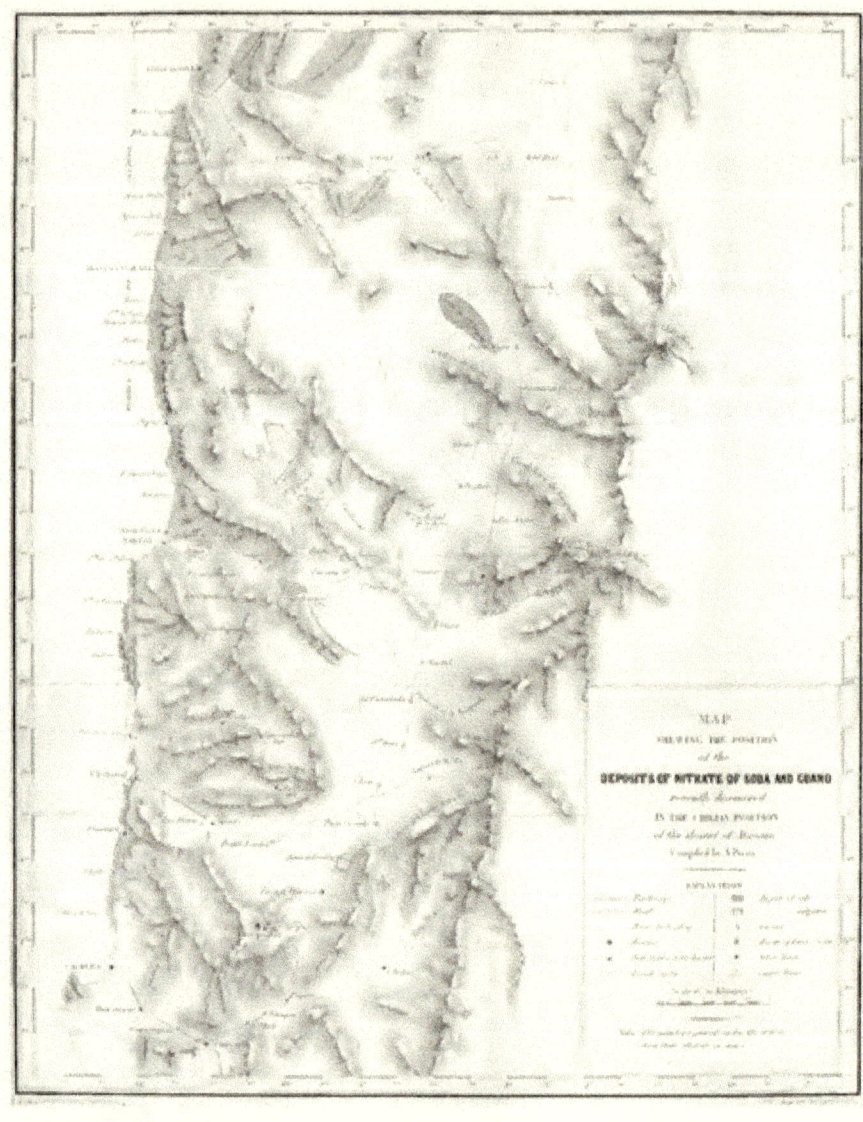

www.ingramcontent.com/pod-product-compliance
Lightning Source LLC
Chambersburg PA
CBHW020331090426
42735CB00009B/1487